About Myths and Facts in the Cardiovascular System of the Giraffe

About Myths and Facts in the Cardiovascular System of the Giraffe

A Morphological Study

PhD Thesis by

Kristine Hovkjær Østergaard

Department of Health Science and Technology,
Aalborg University, Denmark

River Publishers

Aalborg

ISBN 978-87-92982-61-2 (paperback)
ISBN 978-87-92982-60-5 (e-book)

Published, sold and distributed by:
River Publishers
Niels Jernes Vej 10
9220 Aalborg Ø
Denmark

Tel.: +45369953197
www.riverpublishers.com

Preface

The present thesis is based on experimental studies carried out as a part of the Danish Cardiovascular Giraffe Research program (DaGiR). The project is a joint international study involving scientists from health science, biology, engineering, veterinary medicine and also laboratory staff, animal keepers and others. The in vivo studies and tissue sampling were performed during three expeditions to a wildlife quarantine station (WAI) in the Gauteng province, South Africa in 2006, 2010, and 2012.

The project has been on-going for more than six years and has its main focus on the cardiovascular system of the giraffe. Until now, the project has provided new and interesting insight on the fundamental mechanisms of the cardiovascular system. The project seeks to contribute to the continuing scientific work related to cardiovascular diseases, like hypertension, in humans. The project has, with success, investigated how to improve the techniques for safely darting, sedating and anesthetising giraffes, which is most relevant to zoos and wildlife parks.

My work in the present thesis is a part of the DaGiR project and without the entire team of passionate scientists and co-workers the thesis would not have come into existence. Thank you all for that.

I would like to express my deepest gratitude to my main supervisor Professor Ulrik Baandrup; I thank you for the professional, patient and invaluable supervising through all of the work supporting this thesis. Also, my most sincere thanks to you for your support and guidance through hard times, that goes beyond scientific work. Very special thanks to my supervisor professor Tobias Wang; I owe you my deepest gratitude for becoming a part of the DaGiR project and for your dedication to science which is deeply inspiring. Also special and most grateful thanks to my supervisor professor Jens R. Nyengaard; without you I would have been lost right from the beginning in the field of stereology. Thank you for always taking the time to patiently explain and discuss methods and data, and also for hospitality at the department.

I also owe many thanks to Johnnie Bremholm Andersen for the significant amount of time spent on guiding me through the world of hardware and software needed to perform modern stereology.

A lot of people have helped me during these studies. Thanks to everybody at the Center for Clinical Research at Vendsyssel Hospital, in particularly bioanalysts Bente Wormstrup and Mette Skov Mikkelsen who have prepared tremendous amounts of giraffe tissue and also to research secretary Kristina Hansel who can help with answers to almost anything. Huge thanks to everybody at the Stereology and EM laboratory, Aarhus University Hospital for your hospitality and especially to Maj-Britt Lundorf for tissue preparation and Helene Andersen for both scientific and off-topic conversations. Great thanks to everybody at Zoophysiology, Department of Bioscience, Aarhus University for hospitality and for going the distance with me.

Last but not least, huge thanks to Mette and Niels for proofreading and to my family, friends and especially my boyfriend Morten, for support throughout my studies, and for having patience with an absent-minded Ph.D. student.

Contents

List of Papers

Paper I:
Østergaard K.H., Baandrup, U.T., Wang, T., Bertelsen, M.F., Andersen, J.B., Smerup, M. and Neyngaard, J.R. (2013). Left ventricular morphology of the giraffe heart examined by stereological methods. The Anatomical Record. 296; 611–621.

Paper II:
Østergaard, K.H., Bertelsen, M.F., Brøndum, E.T., Aalkjær, C., Hasenkam, J.M., Smerup, M., Wang, T. and Baandrup, U.T. (2011). Pressure profile and structural changes of the arteries along the giraffe hind limb. Journal of Comparative Physiology B. 181; 691–698.

Paper III:
Petersen, K.K., Hørlyck, A., Østergaard, K.H., Andresen, J., Broegger, T., Skovgaard, N., Telinius, N., Laher, I., Bertelsen, M.F., Grøndahl, C., Smerup, M., Secher, N.H., Brøndum, E.T., Hasenkam, J.M., Wang, T., Baandrup, U and Aalkjær, C. (2013). Protection against high intravascular pressure in the leg of giraffes. Submitted to American Journal of Physiology - Regulatory, Integrative and Comparative Physiology

Paper IV:
Brøndum, E.T., Hasenkam, J.M., Secher, N.H., Bertelsen, M.F., Grøndahl, C., Petersen, K.K., Buhl, R., Aalkjær, C., Baandrup, U.T., Nygaard, H., Smerup, M., Stegmann, G.F., Sloth, E., Østergaard, K.H., Nissen, P., Runge, M., Pitsillides, K. and Wang, T. (2009). Jugular venous pooling during lowering of the head affects blood pressure of the anesthetised giraffe. American Journal of Physiology 297:1058–1065.

Abrevations
CVtot, coefficient of variation
DaGiR, Danish cardiovascular Giraffe Research Program
IUR, isotropic uniform random sampling
LV, left ventricle

MAP, mean arterial pressure
PAS-M, periodic Schiff-sampling Methenamine Silver
PBS, phosphate buffered saline
SD, standard deviation
SURS, systematic uniform random
UR, uniform random sampling

Background

The giraffe can reach a height of up to 6 meters. The long neck of the giraffe (*Giraffacamelopardalis*) has probably evolved as an adaptation to browse for food no other mammal can reach in combination with sexual selection for the highest and strongest males (Simmons and Altwegg, 2010). Irrespective of the selective pressure, the long neck poses serious challenges to the cardiovascular system of the giraffe.

By virtue of being the tallest living animal on earth, giraffes are endowed with a mean arterial blood pressure (MAP) twice as high as other mammals (Figure 1) (Goetz and Budtz-Olsen, 1955; Van Citters et al, 1966; Mitchell et al, 2006; Brondum et al, 2009). In addition,giraffes also experience larger hydrostatic pressure differences within their cardiovascular system than any other animal (Fig. 1) (Ostergaard et al, 2011). The high MAP is probably necessary to ensure adequate brain perfusion as the head is situated more than two meters above heart level (Brondum et al, 2009). In addition, it has been suggested that cerebral perfusion is supported by a siphon mechanism, where the gravity in the carotid artery is overcome by energy recovered as blood returns to the heart through the jugular vein (Hicks and Badeer, 1989; Badeer and Hicks, 1992; Pedley et al, 1996; Brook and Pedley, 2002).

In humans, a high MAP is associated with health related problems like altered structures of small resistance arteries (Korsgaard et al, 1993; Mulvany, 2002), atherosclerosis, heart and kidney diseases etc. (Kumar et al, 2005; Barri, 2008). In giraffes, mechanisms probably have evolved to protect the cardio-vascular system from the adverse problems of a high MAP. The giraffe heart experiences conditions that would be regarded extreme to humans and other mammals. Already in the 1950ies, it was proposed that giraffes do not develop heart failure because the relative mass of their heart is almost twice that of other mammals (Goetz and Keen, 1957). However, recent studies have shown that the mass of the giraffe heart constitutes 0.5% of body mass (Mitchell and Skinner, 2009; Brondum et al, 2009), just as in other mammals, including humans.

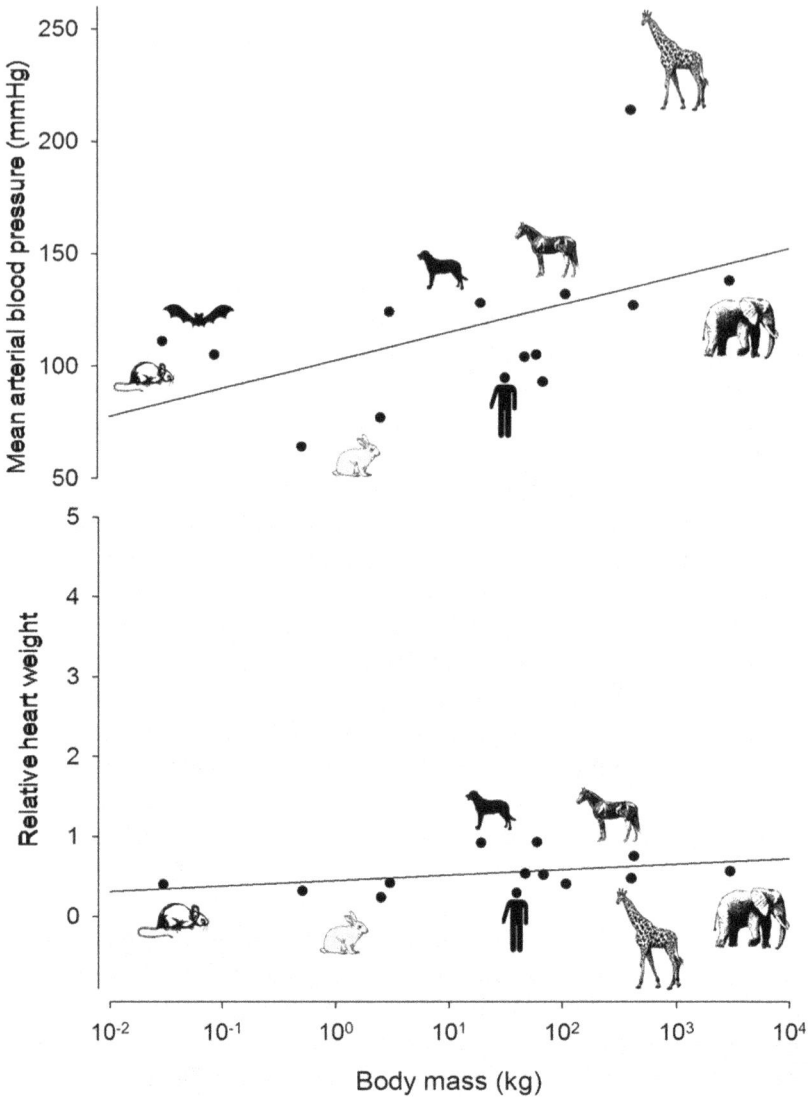

Figure 1 While the mean arterial pressure (MAP) of the giraffe is twice as high as that of other mammals including humans (A), the relative heart mass, is similar (B). All data except for the giraffe are from Seymour and Blaylock, (2000). (Original figure in Paper I).

Not only the heart, but also the systemic circulation must be seriously challenged by the high MAP. As the arterial pressure in the hoof is the sum of the MAP and the hydrostatic pressure from a blood column of approximately

2 meters (difference between the heart and the hoof), the arterial pressure at the hoof approximates 350 mmHg (Ostergaard et al, 2011). Mechanisms to reduce problems of a distal pressure that high probably have evolved to prevent damages like oedema in the giraffe legs (Goetz and Keen, 1957; Goetz et al, 1960; Williams et al, 1971; Hargens et al, 1987; Kimani et al, 1991).

Primarily by studying morphology, we aim toto understand the adaptations and mechanisms underlying function and protection of the cardiovascular system exposed to the highest known pressure of any present-day mammals. We seek to confirm or reject some of the myths that inevitably exist in connection with a mammal reaching 5–6 meters in height.

DaGiR

The Danish cardiovascular Giraffe Research program (DaGiR) was established in 2006 as an ambitious large-scale collaborative study of the giraffe cardiovascular system involving scientists from health science, biology, veterinary medicine and engineering as well as several co-workers. Since then, DaGiR orchestrated three expeditions to The Gauteng province (South Africa) to investigate giraffe cardiovascular system. The studies have had several parallel setups, including; 1) measurements of pressure and flow in free-ranging giraffes by wireless monitoring of implanted probes, 2) acute measurements of physiological parameters of anesthetised giraffes, 3) *in vitro* and *post mortem* studies of heart, arteries and veins, kidney etc. In addition, tissue has been collected from zoo giraffes, including neonates and calves that were euthanized for various reasons.

Morphological Study of the Giraffe Cardiovascular System

With this thesis, we correlate morphology with functional parameters of the giraffe cardiovascular system. More specifically, the focus has been on the heart, the conduit arteries of the lower extremities, the neck and abdominal large veins (jugular and inferior caval vein) and carotid artery. As the giraffe heart experiences a pressure load twice as high as other mammals, we conducted a quantitative morphological study to gain knowledge of the possible adaptations of the giraffe left ventricle (*Paper I: Left ventricular morphology of the giraffe heart examined by stereological methods, Ostergaard et al, 2013*). This is a detailed stereological study of numbers, densities and volumes of cardiomyocytes, non cardiomyocytes and capillaries of the giraffe left ventricle. These morphological characters provide knowledge of how the "normal sized" giraffe heart works against a high pressure gradient and enable us to relate the results to other mammalian hearts. The relative heart mass was determined to be 0.5% of body mass, which is similar to other mammals. A significantly higher number of nuclei in the adult myocardium compared to the young giraffe myocardium was found. Also an unusually high number of nuclei per cardiomyocyte was measured, compared to any other mammalian myocyte that we know of. These findings are discussed in *paper I*.

The lower part of the vascular system of the giraffe is exposed to a large hydrostatic pressure and the effect of gravity is significant. In humans, hypertension causes arterial remodelling, such as increase of the media/lumen ratio. These alterations may have different characteristics and origins (Fig. 2) (Mulvany, 1998). Also the elastin/collagen ratio of the arterial wall may increase and affect the elastic modulus, the arteries' ability to reform and recoil. A stereological analysis of the conduit arteries of the giraffe hindleg was conducted along with a pressure profile from the legs (*Paper II: Pressure profile and morphology of the arteries along the hindlimb of the giraffe, (Ostergaard et al, 2011)*. The lumen area and tunica media area were measured and the fraction of elastin in the arterial wall were estimated. These results revealed a sudden sphincter-like constriction of the tibial artery, which was suprising as the tibial artery is a large conduit artery. The innervation density was significantly higher in the area of the constriction. Arterial and venous pressure profiles were generated in the forelegs of four giraffes, revealing a mean arterial pressure drop not significantly different from what would be expected from gravity. The pressure profile enabled us to relate the morphological findings.

The findings of a sphincter-like structure in the conduit arteries was followed up by ultrasound studies of the these arteries (*Paper III: Protection against high intravascular pressure in the leg of the giraffes, Petersen et al, submitted*).

The ultrasound examinations together with a specific immunohistochemical staining (Tyroxine hydroxylase) of the innervation confirmed the existence of a sphincter-like structure in both the foreleg (median artery) and hindleg (tibial artery). The ultrasound imaging also demonstrated spontaneous contractions in this area of the artery. Pressure and flow measurements from just proximal to distal of the sphincter area were undertaken. The results are discussed in *paper III*.

The regulation of blood pressure and flow to the brain of the giraffes has been a "hot topic" for centuries. Especially two theories have been debated, the one being "the siphon mechanism" (Hicks and Badeer, 1989; Badeer and Hicks, 1992; Pedley et al, 1996; Brook and Pedley, 2002) and the other one "the

hypotrophic   eutrophic   hypertrophic

inward

outward

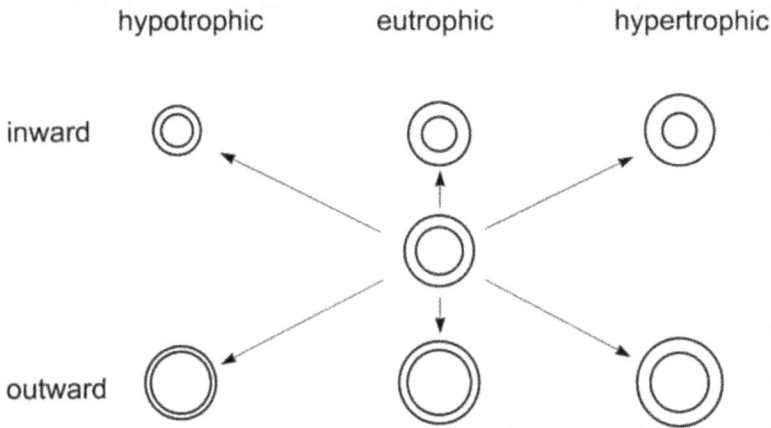

Figure 2 The remodeling of arteries origin from different events and may have different character. Here, the center vessel is the starting point, the "normal" vessel. Hypotrophic remodeling (left vessels in the figure) is characterized by a decrease of the cross sectional area of the arterial wall, either inward or outward (meaning with either a decrease or increase in lumen diameter, respectively). In eutrophic remodeling no change in cross sectional area of the wall takes place; the eutrophic remodeling can be both inward and outward (mid upper and lower vessel in the figure). Hypertrophic remodeling is characterized by an increase in cross sectional area either inward or outward (Right vessels in the figure). *Mulvany 1998.*

waterfall analogy" (Pedley et al, 1996). To approach the questions concerning how the cerebral circulation is regulated, blood pressure, flow and cross sectional area of the carotid artery and jugular vein were simultanously measured as well as the heart rate (*Paper IV: Jugular venous pooling during lowering of the head affects blood pressure of the anesthetized giraffe (Brondum et al, 2009)*). Besides the heamodynamic measurements, macroscopic examinations of the carotid artery, jugular vein and the heart were carried out. There seems to have been an unproven accepted belief that the carotid artery of the giraffe had valves, mentioned in literature such as Encyclopedia Britanica online and The Danish Encyclopedia (2009). However, no valves were present. Moreover, the general belief that the relative mass of the giraffe heart is bigger than other mammalian hearts was proven wrong. The giraffe heart had the relative same mass as any other mammalian heart, including the human heart. The primary findings revealed that the cerebral circulation is maintained by a high arterial pressure and not by a siphon mechanism or waterfall analogy. This is further discussed in *paper IV.*

Materials and Methods

Tissue Collecting

During the three expeditions, heart, blood vessels and other tissues were collected for morphological studies. A total of 34 giraffes were handled in South Africa during the three expeditions. Before euthanasia and autopsy, physiological measurements were performed either under anesthesia or in free-ranging animals (Fig. 3). Not all of these physiological experiments are described here. All giraffes were between 2 and 5 years old with a body mass between 380–654 kg and a height between 3.1–3.9 m. The animal handling procedure and ethic approvals were described in details in paper IV (2006 expedition). In addition, tissues were collected from giraffes (adults, youngs and new borns/still borns) from different zoos, the animals being euthanized for various health related reasons. As the studies performed at each expedition vary in focus and methods, the number of giraffes, tissue samples, measurement etc. used for each study varies (Table 1).

Stereology–Principles and Probes

Stereology is based on geometry and statistics. It is a method to obtain quantitative unbiased information of a three-dimensional material, but from measurements made on two-dimensional section planes (Gundersen et al, 1988;

Table 1 Overview of Tissue Collected at the Three Expeditions and from Various Zoos

Expedition	Giraffes	Age	Hearts	Arteries (Lower Extremeties)	Jugular Veins	Carotid Arteries	Inferior Caval Vein
2006	7	adult	6	14	11	12	–
2010	17	adult	11	19	–	–	–
2012	10	adult	8	–	7	–	7
zoo giraffes	6	adult	3	1	2	–	–
	3	neonate	2	2	–	–	–
	3	young	2	1	4	–	–
Total	**44**		**32**	**37**	**24**	**12**	**7**

Nyengaard, 1999). Using the appropriate stereological tools you can avoid the limitations of two-dimensional methods. One of the major limitations is that two-dimensional methods make assumptions of the three-dimensional structure of the tissue or organ under consideration (Gundersen et al, 1988). Avoiding these assumptions is highly desirable; since they provide imprecise results. By ensuring random position and random orientation when you sample your tissue, the probability of hitting a specific structure with various probes is proportional to the number or dimension of that structure. Random position is ensured by making systematic uniform random sampling (SURS). The basic principles are the same as with a uniform random sampling (UR). However by sectioning the tissue in slices with a fixed distance (systematic) it will reduce

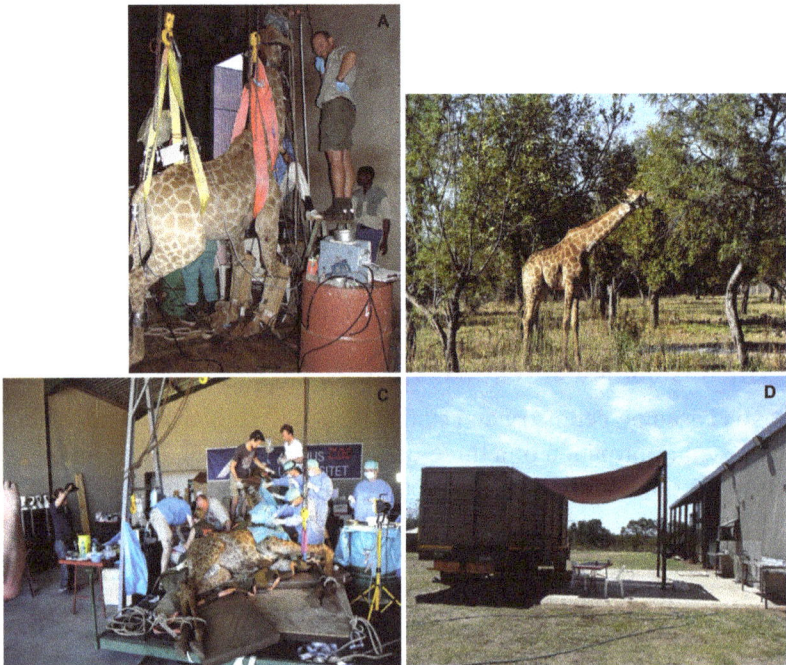

Figure 3 Experimental setup in South Africa. Picture **A** shows the setup from the 2006 expedition. The anestitized giraffes were hoisted upright in a "slings" situated in a stable, while functional parameters as different pressures, flow, heart rate etc, were measured. Picture **B** show the setup from 2012, where the giraffes were put on an elevated table, situated in a large hangar with a hoist system. Picture **C** is an implanted free ranging giraffe being monitored, note the shaved area on the lower neck where the implantation was performed. Picture **D** is the postmortem site where the autopsy and tissue collection were carried out. (Pictures: Rasmus Buchanan and Kristine Østergaard).

the slices needed to equally represent the whole heart. By positioning the first cut randomly you will still ensure randomness (Gundersen and Jensen, 1987). Further subsampling should also follow the SURS principle and this way ensuring an accurate and precise estimate of the structure of interest. Random orientation of the tissue may be ensured by isotropic uniform random sampling (IUR). We created IUR sections using the isector (Nyengaard and Gundersen, 1992) embedding the tissue in small agar spheres (Figure 3). By making IUR sampling it is possible to estimate all stereological quantities like number, volume, surface area, length, spatial distribution, connectivity, shape analysis etc.

Not all parameters need isotropic sampling. For example number and global volume is independent of orientation, whereas length and surface area are affected by the orientation of sections. The use of IUR sections makes it possible to generate isotropic test planes to be used for estimation of length and surface area. In contrast, vertical uniform random (VUR) sections can only generate isotropic test lines, which can be used for surface area estimation but not length estimation on thin sections.

The correct stereological probe to measure specific parameters can be determined by the following rule:

$$D(f) + D(p) \geq 3$$

The dimensions of the features you wish to study, $D(f)$, together with the dimensions of the probe, $D(p)$ has to be equal to or greater than 3. For example, volume is a three-dimensional feature, and can be estimated with test points which have no dimensions, or surface area which have two dimensions and can be estimated using test lines having one dimension.

Reference volume and reference trap

The Cavalieri estimator is efficient to measure total volumes (reference volume) of an organ or specific region of an organ, no matter how irregular the shape of the structure is (Gundersen and Jensen, 1987):

$$V(LV) = \bar{t} \cdot (a/p) \cdot \sum P(LV) \qquad (1)$$

$V(LV)$ is the total volume of the left ventricle, LV, $\bar{t}$ is the average distance between sampled slices, (a/p) is the area associated with each test point in the counting grid and $\sum P(LV)$ is the total number of points hitting the LV. When measuring densities of numbers, volume, surface area, length etc., the reference volume is used to convert these to total quantities to avoid the reference trap (Brændgaard and Gundersen, 1986). The reference trap is referred to when wrong conclusions are drawn from density measurements alone, and this is the reason the whole organ is required, thus the reference volume can be estimated. The Cavalieri estimator does not need isotropic tissue samples, however the method is affected by deformation of tissue (Brændgaard and Gundersen, 1986).

Tissue shrinkage

When measuring volumes and densities it is important to consider tissue deformation, such as shrinkage, which can have great influence on measurements. To avoid shrinkage of the specimens used for measuring volumes and densities, plastic (glycolmethacrylate) embedment was performed (Dorph-Petersen et al, 2001). Shrinkage was estimated by determining the tissue volume before and after embedding. Before embedding the volume was determined by transforming the wet weight of the tissue into volume. After embedding in plastic the volume was determined using the Cavalieri estimator (Gundersen and Jensen, 1987). Mean volume shrinkage was $6 \pm 0.2\%$, which did not differ significantly from 0 and no correction for shrinkage was made.

Heart Morphology

Tissue preparation

21 hearts from adult giraffes, two from still-borns, two from young animals (4–6 months old) and one from a foetuswere collected for morphological studies. After euthanasia, the hearts were weighed and immersed in 4% phosphate buffered formaldehyde for 24–48 h and then stored in phosphate buffered saline (PBS) with azid (antifungal compound). The left ventricles (LV), including the septum, were isolated and weighed and the following tissue sampling was conducted by the principles of systematic, uniform and random sampling (SURS) (Gundersen and Jensen, 1987). To ensure isotropic orientation of the tissue blocks, the isector was used to embed every small tissue block in a small agar sphere (Fig. 4) (Nyengaard and Gundersen, 1992) before embedding the samples in paraffin or plastic (glycolmethacrylate). For the different measurements, one 30 μm and two 3 μm thick slices were cut from each block and further series of 16 slices 2 μm thick were sectioned from each block. The 30 μm slices were stained with Mayer's heamatoxylin (Bie and Berntsen) for counting nuclei. One set of 3 μm slices was stained with Mayer's heamatoxylin and 0.15% basic fuchsin for myocyte volume count and one set

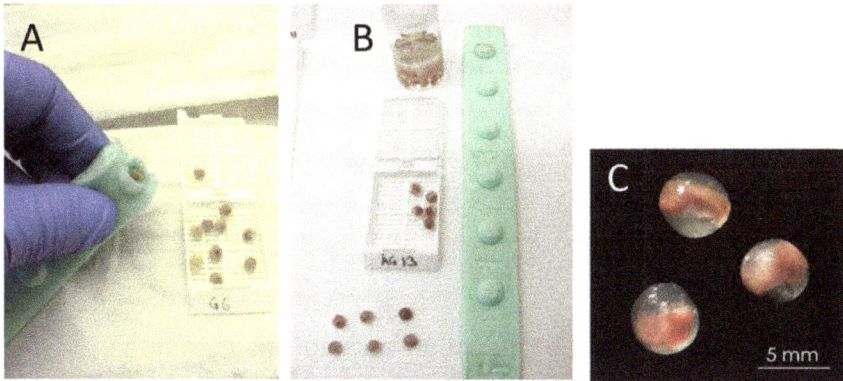

Figure 4 To create isotropic uniform random tissue (IUR) sections, the samples were embedded in isectors (**A, B, C**). I.e. the tissue blocks are embedded in small agar spheres (**C**) using a rubber mould (**A, B**). Before embedment in paraffin or plastic they were rolled on the table to ensure random orientation.

was stained with polyclonal antibody CD31 (Abcam®) for capillary volume estimation. The serial slices were stained with periodic Schiff-Methenamine Silver (PAS-M), which stains the intercalated discs of the cardiomyocytes enabling us to differentiate them, and thereby count the nuclei per myocyte.

Stereological counting equipment

For analysis and counting an Olympus BX51 microscope with a motorized stage (Prior) with an insert of eight slices was used. An electronic microcator (Heidenhain) measuring the z-axis thickness of the tissue was connected to the microscope and to a personal computer as was also a digital camera (Olumpus DP70). For superimposing of the stereological probes onto the tissue and for counting and analysis, the computer was equipped with newCAST stereology software (Visiopharm, Hørsholm, Denmark).

Heart mass, volume fraction and total volume of myocytes and non-myocytes

Heart weight was determined and left ventricular volume (reference volume) was estimated using the Cavalieri estimator (equation 1) (Gundersen and Jensen, 1987). Using point counting on 3 μm plastic sections the volume fraction of cardiomyocytes and non cardiomyocytes (anything else than myocytes) was estimated using equation 2:

$$Vv \frac{Myocytes}{LV} = \frac{\sum P(myocyte)}{\sum P(LV)} \qquad (2)$$

Volume estimated using point grids is independent of tissue orientation, but sensitive to tissue shrinkage and swelling.

Total volume was estimated by multiplying the volume fraction with the left ventricular volume i.e. the reference volume.

Density and total number of myocyte- and non-mycyte nuclei

The numerical density of myocyte nuclei and non-myocyte nuclei were estimated using the optical disector. An unbiased counting frame with a known area was superimposed onto the 30 μm thick sections. The counting frame and the known z-axis height ensure that we know the volume of the disector. A generated z-axis distribution of nucleishowed that the distribution of nuclei was uniform in the 30 μm thick sections except a few μm at the top and bottom. A guard height of >5 μm was established at the top and bottom of the sections and the 15 μm high disector was placed in the centre of the thick sections. The microscope was focused down through the sections and when a nucleus was in focus and sampled by the counting frame, it was counted (Fig. 5) (for detailed description see paper I). The numerical density of myocyte nucleiNv(myocyte nuclei/LV) and non-myocyte nuclei (Nv(non-myocyte nuclei/LV) were obtained using equation 3:

$$Nv(myocyte\ nuclei/LV) =$$

$$\frac{\frac{\sum (t_i q_i^-(myocyte\ nuclei))}{\sum (t_i q_i^-(myocyte\ nuclei))}}{MA} \cdot \frac{\sum Q^-(t_i q_i(myocyte\ nuclei))}{\sum P(LV)} \cdot \left[\frac{p}{h\,a(frame)} \right] \qquad (3)$$

t_i is the local height of the section at position i, q_i^- is sampled number of myocyte nuclei at position I and i is the i'th field of view. MA is the microtome advance, $\sum Q^-$ is total number of myocyte nuclei counted in all disectors of one LV. $\sum P(LV)$ is the total number of points falling on ventricular tissue (the reference space), p is the number of counting frame corner points used to count the reference space, and h is height of the disector and a is the area of the counting frame. From these data, also the total number of myocyte and non-myocytes were determined by multiplying the numerical densities by the LV volume (reference volume).

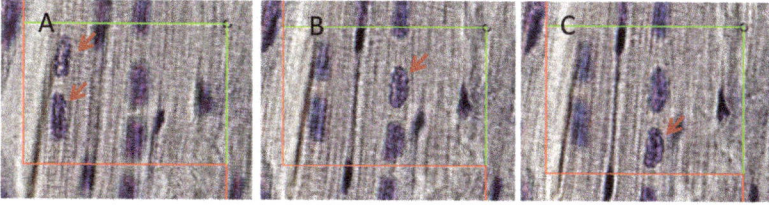

Figure 5 Using the optical disector, a counting frame and 30 μm thick tissue sections, the density of nuclei was estimated. The picture illustrate when the microscope is focused down through the sampling volume. A nucleus is counted when it is in focus, within the counting frame, not touching the exclusion lines of the counting frame (red line). **A:** two nuclei are in focus and counted (arrows). **B:** the two nuclei from picture A is no longer in focus, but a third nucleus is now in focus (arrow) and is counted. **C:** The first three nuclei are now out of focus and a fourth in focus (arrow) and this is counted too.

Mean number of nuclei per myocyte

The number of nuclei per myocyte was estimated using the physical disector. The 16 serial cut sections stained with PAS-M were mounted on eight slides which were mounted on the motorized stage in the microscope and using the newCAST software they were delineated. In this way, the same regions of interest were followed throughout the 16 serial cut sections. As the tissue was isotropic, to ensure that all orientations occur with equal likelihood, we could freely choose field of view where the myocytes were longitudinally cut. These fields of view are called "vertical windows". All counts were started with section seven or eight as reference sections. An unbiased counting frame was superimposed onto the tissue. When a nucleus was present in the reference section but not in the next "look up" section, it was counted and by following this myocyte profile both up and down in the section series until they disappeared, all nuclei for this particular myocyte were counted. The mean number of nuclei was determined from equation 4: N(nuc/myocyte)=

$$\frac{\sum Q^-(mono) + Q^-(bi) + Q^-(tri) + Q^-(quattro)}{Q^-(five) + Q^-(six) + Q^-(seven) + Q^-(eight)}$$
$$\frac{+\dfrac{\sum Q^-(mono)}{1} + \dfrac{\sum Q^-(bi)}{2} + \dfrac{\sum Q^-(tri)}{3} + \dfrac{\sum Q^-(quattro)}{4}}{+\dfrac{\sum Q^-(five)}{5} + \dfrac{\sum Q^-(six)}{6} + \dfrac{\sum Q^-(seven)}{7} + \dfrac{\sum Q^-(eight)}{8}} \quad (4)$$

Where Q^- (mono) is the number of myocytes with one nucleus, Q^- (bi) is the number of myocytes with two nuclei, Q^- (tri) is the number of myocytes with tree nuclei and so forth up to Q^- (eight), which is the number of myocytes with eight nuclei. Myocytes with two or more nuclei had a larger probability of being sampled than myocytes with one nucleus, to eliminate the effect of this, the number of myocytes with one nucleus was divided by one, myocytes with two nulcei was divided by two etc.

Total number and density of myocytes and mean myocyte volume

The total number of myocytes in the LV was estimated by dividing the total number of myocyte nuclei with the average number of nuclei per myocyte. The numerical density of myocytes was estimated by dividing total number of myocytes with the total volume of the LV, the reference volume. Further, the mean volume per myocyte was estimated by dividing the total volume of myocytes with the total number of myocyte in the LV.

Capillary volume fraction and total volume

The volume fraction of capillaries was estimated using a point counting grid on 3 μm thick sections stained with a polyclonal antibody against CD31. The volume fraction was estimated using formula 5:

$$Vv(\frac{capillary}{LV}) = \frac{\sum P(capillary)}{\sum P(LV)} \tag{5}$$

$\sum$P (capillary) is the total number of point hitting capillaries in the LV and $\sum$P(LV) is the total point hitting the LV. The total volume of capillaries was estimated by multiplying the volume fraction of capillaries with the volume of the LV (reference volume).

Arterial Morphology

Tissue preparation

For the histological examinations of the large arteries of the lower extremities 29 tibial arteries from 19 giraffes and 7 median arteries from 7 giraffes were investigated. The studies of the arterial morphology was extended over two expeditions (2006 and 2012) and as the first gave valuable knowledge and inspiration, the second expedition used different procedures for tissue collection, based on the experience gained from the first expedition. 19 femoral/tibial arteries from the hindleg from 10 giraffes were obtained (2006). For each 15 cm of the artery a tissue block was cut for microscopic investigation. In addition, five tissue blocks were collected from the area where the narrowing of the artery was observed. All tissue blocks were fixed in 4% phosphate-buffered formaldehyde. The tissue was prepared for microscopy and cut in 3 μm thick sections and stained with S100 antibody (Dako), Mayer's heamatoxylin (Bie and Berntsen) and Van Gieson (local recipe), respectively. In 2012, a total of 7 median and 10 tibial arteries were collected. The now known sphincter area was located, and 8 tissue blocks were sampled from this area. All blocks were fixed in 4% phosphate-buffered formaldehyde for 24–48 h and stored in PBS + azid. The tissue was prepared for microscopy and 3 μm thick sections were cut. One section from each block was stained with S100 antibody and one section from each was stained with Tyrosine-Hydroxylase (TH) (Abcam®). The investigation was conducted with an Olympus MVX10 macrofluorescence and stereo microscope system.

Area, volume fraction of elastin and innervation

Using the 2D nucleator from the newCAST software the cross sectional area of lumen and tunica media of the femoral/tibial artery was determined. The 2D nucleator is an efficient tool for measuring areas at the local level, and does not need information of reference volume or disorientation of sections (Gundersen, 1988). With point count probes the volume fraction of elastin and the innervation of the arteries was estimated (for detailed descriptions see paper II and IV).

Carotid artery

12 carotid arteries were obtained from 7 giraffes (2006). The arteries were cleansed from excess tissue and cut open longitudinally and the presence of arterial valves was examined

Pressure Measurements

An arterial and venous pressure profile along the length of the foreleg was measured in four giraffes under anesthesia by introducing a tip pressure catheter through a sheath 30 cm above the hoof and then advanced in steps towards the heart (For details see paper II). The pressure difference across the sphincter-like area of the median artery was measured in 7 giraffes by introducing a catheter just proximal to the sphincter and a catheter 20 cm proximal to the hoof (For details see paper IV).

Table 2 Locations from Where the Tissue Samples were Collected from the Caval and Jugular Vein

Inferior Caval Vein	Jugula Vein
1 cm proximal the iliaca vein	2 cm proximal R. Atrium
1 cm proximal the renal vein	As rostral as possible
3 cm distal R. Atrium	

Morphology of the Inferior Caval Vein and Jugular Vein

The jugular veins from 7 giraffes were cut open and the number of and distance between valves was counted (2006). Samples from predetermined locations (Table 2) of the jugular vein and inferior caval vein were obtained from 8 giraffes (2012). Samples were fixed in 4% phosphate-buffered formaldehyde and then stored in PBS + azid. The tissue was prepared for microscopy and 3 μm sections were cut and stained with Masson's tricrome (local recipe). An Olympus MVX10 macrofluorescence and stereo microscope system was used for examination and the newCAST software was used to count and analyze the volume fraction of muscle tissue (for detailed description see paper V).

Statistics

All heart data are presented as mean $\pm$ SD (standard deviation) in the figures and in the text data are presented as mean (CVtot). CVtot is the coefficient of variation (SD/mean). All arterial and venous data is presented as mean $\pm$ SD and pseudo replication is avoided by calculating mean for each giraffe before calculating for all giraffes. Differences are tested using one-way analysis of variance (ANOVA).

Results

Heart

Heart mass, volume fraction and total volume of myocytes and non-myocytes

The relative heart mass of the adult giraffe heart was 0.5%. The volume fraction of myocytes in the LV was as expected; and constituted the same fraction in adult and young LV (0.89 and 0.90, respectively). The volume fraction of non-myocytes was 0.1 in both adult and young LV. The total volume of myocytes in the adult heart was 1153 cm^3 whereas it was 68 cm^3 and 119 cm^3 in the still borns and 4–6 months animals, respectively. Also the non-myocyte total volume was estimated and the relative differences between ages were the same as for the myocytes (Table 3).

Density, total volume of myocyte- and non-mycyte nuclei and nuclei per myocyte

The numerical density of nuclei was several times higher in the young hearts compare to adult hearts (Table 3). More surprisingly, the total number of myocyte nuclei was 2.6 times higher in the adult heart than in the young heart. The number of nuclei per myocyte ranged between 2 and 8 eight nuclei (Fig. 6) and the mean number of nuclei per myocyte was 4.2.

Total volume and density of myocytes and mean myocyte volume

The total number of myocytes was 2.7×10^{10} and the numerical density was 24.1×10^6 cm^{-3} in the adult giraffe LV. Unfortunately, it was impossible to retrieve these data from the young LV, as the young LV tissue did not stain well. The average volume of a myocyte from the adult LV was 39.5×10^3 μm^3 (Table 3).

Table 3 Volume, Densities and Numerical Data from the Giraffe Left Ventricle (Original
Figure: Paper II)

	Young	VCtot	Adult	VCtot
LV volume, cm^3	104	0.31	1312	0.3
Vv (myocyte/LV)	0.9	0.01	0.89	0.02
Vv (non-myocyte/LV)	0.1	0.19	0.11	0.9
Vv (myocyte/LV), cm^3	93.6	0.32	1153	0.28
V (non-myocyte/LV), cm^3	10.7	0.27	136	0.3
Mean myocyte volyme, um^3			39508	0.26
N (myocyte/LV)			$2.7 \cdot 10^{10}$	0.25
Nv (myocyte nuclei/LV), mm^{-3}	$504 \cdot 10^3$	0.17	$120 \cdot 10^3$	0.2
Nv (non-myocyte nuclei/LV), mm^{-3}	$72 \cdot 10$	0.16	$74 \cdot 10^3$	0.21
N (myocyte nuclei)	$4.9 \cdot 10^{10}$	0.26	$1.3 \cdot 10^{11}$	0.21
N (non-myocyte nuclei)	$5.9 \cdot 10^9$	0.5	$7.7 \cdot 10^{10}$	0.2
N nuclei per myocyte			4.2	0.07

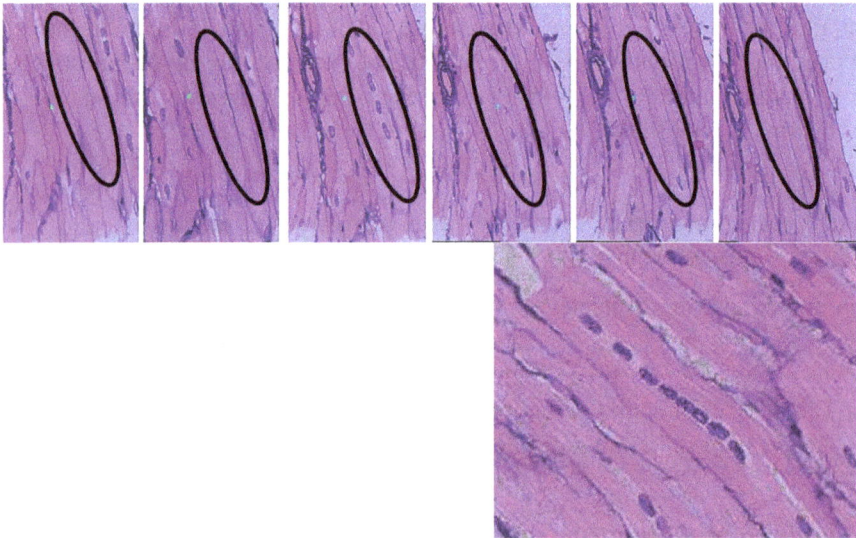

Figure 6 6 of 16 serial cut sections from the physical disector following a myocyte with four
nuclei through the 6 sections. B: myocyte with 8 nuclei.

Capillary volume fraction and total volume

The mean capillary volume fraction of the adult LV was 0.054 and total volume
of capillaries was 65.6 cm^3.

Arteries

The carotid artery

We did not find any valves in the carotid artery of the seven animals studied.

Area of the median–and tibial arteries

A sudden large decrease of the lumen of the femoral/tibial artery was observed. The total area of lumen and media decreased 80% from top to bottom. 43% of this decrease occurred within 2–4 cm just below knee height (Fig. 7 and 8). 54.5 % of the total lumen area decrease happened within these 2–4 cm and 85 % of the media:lumen area ratio decrease occurred in this area just below

Figure 7 Left: A longitudinal cut of the tibial artery just across the sudden narrowing of the artery. 85% of the change in media to lumen area along the artery happened within this area of 2–4 cm. The black arrows pinpoints the lumen above and below the narrowing. Right: indicate the location of the narrowing.

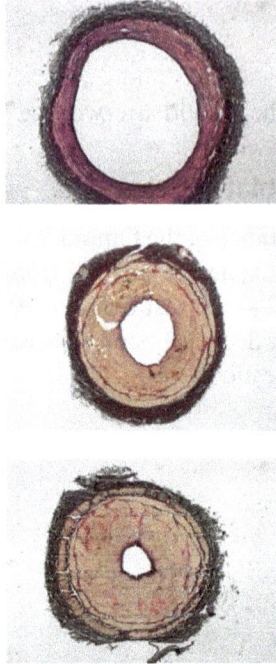

Figure 8 Three sections from the site of the sudden narrowing of the tibial artery cut with a distance of one centimeters between each. The changing media to lumen ratio is obvious. The elastic Van Gieson staining shows the decrease in elastin content of the tunica media and the large increase in muscular tissue from above to below the narrowing.

knee height (Fig. 7 and 8). The same sudden decrease was observed for the media area.

Volume fraction of elastin

The volume fraction of elastin in the femoral/tibial artery was 38% proximal to the narrowing whereas it was 5.8% distally. 59% of this decrease occurred at the site of narrowing (Fig. 8).

Innervation

The S100 staining of the femoral/tibial artery revealed a significantly higher innervation at the area of the narrowing, compared to both proximal and distal to this area. The nerves were located in the adventitia, but some nerves were observed penetrating the media at the site of narrowing.

Calf artery

The same sudden narrowing of the femoral/tibial artery was observed in one neonate and one six months old calf. In a third neonate, the change was less obvious, though still present.

Pressure Measurements

The mean arterial pressure increased 60 mmHg along the foreleg. Proximal in the foreleg the mean arterial pressure was 232 ± 58 mmHg and distally the mean arterial pressure was 292 ± 49 mmHg (n: 4). The relation between the pressure increase and distance did not differ from that predicted by gravity (Fig. 9). To avoid bias due to changes in MAP, the pressure was corrected for MAP (Fig. 9b). The highest arterial pressure measured was 350 mmHg in the distal part of the foreleg of a 322 cm tall giraffe. The mean venous pressure 30 ± 22 mmHg in the proximal part and 113 ± 39 mmHg in the distal part, which was also in accordance with gravity.

Inferior Caval Vein and Jugular Vein

The cross sectional area of the inferior cavalvein differs significantly along its length (Fig. 10). Especially 1 cm proximal to the renal vein, the wall of the caval vein was very thick and the cross sectional area was 224 mm^2 compared to 172 mm^2 and 188 mm^2 near the atrium and near the iliac vein, respectively. The volume fraction of muscle tissue was significantly higher 1 cm above the renal vein (0.62 ± 0.05) compared to the areas near the right atrium (0.26 ± 0.04) and iliac vein (0.39 ± 0.1).

In the jugular vein the volume fraction of muscle tissue was significantly higher in the rostral region of the vein, near the head (0.32 ± 0.02), compared to the volume fraction of in the lower region near the heart (0.17 ± 0.04) (Fig. 11).

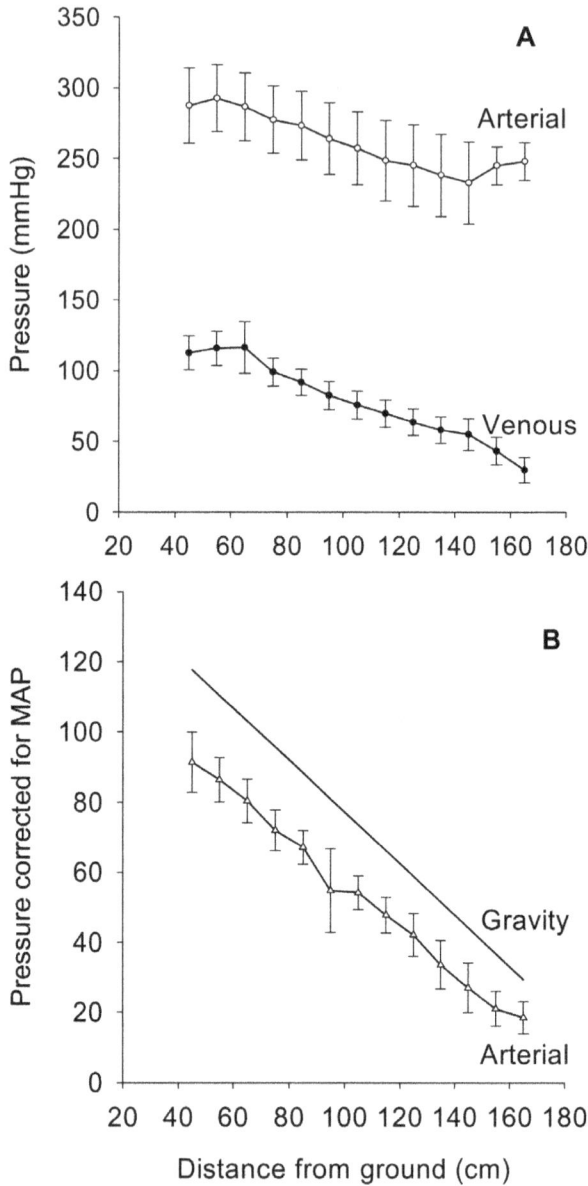

Figure 9 Pressure profile from the giraffe foreleg. **A** arterial and venous pressure profile. The x-axis shows the distance from ground level. **B** The arterial pressure corrected for mean arterial pressure (MAP) is plotted to avoid bias from changes in MAP during measuring. The line representing gravity is also plotted. Data are presented as mean $\pm$ SD n = 4.

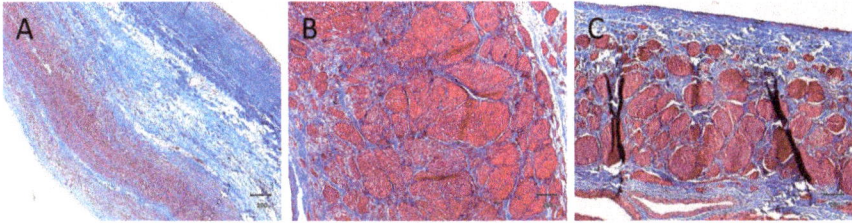

Figure 10 Caval vein morphology. The samples are stained with Masson´s tricrome and illustrate the local large difference in volume fraction of muscular tissue (red staining). **A**: Three cm distal to the right atrium, **B**: 1 cm proximal to the renal vein. **C**: 1 cm proximal to the iliac vein.

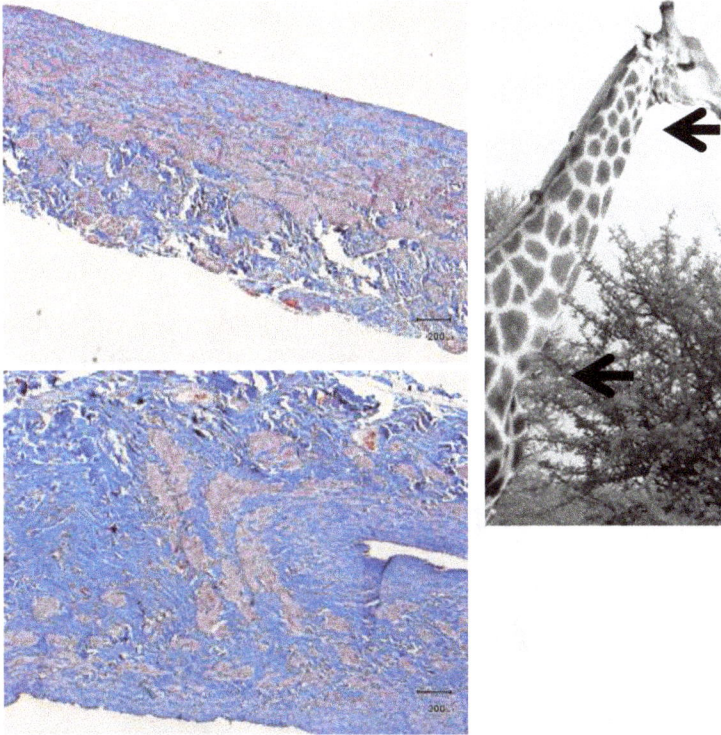

Figure 11 Jugular vein morphology. The upper picture shows a tissue sample from the upper neck (location is pointed out by the upper arrow). The lower picture is a sample from the lower neck, near the heart (location pointed out by the lower arrow). The tissue is stained with Masson´s tricrome and illustrates the higher volume fraction of muscular tissue (light red staining) in the upper region of the jugular veins.

Discussion

That the giraffe heart is larger than other mammalian hearts seems to have been a general belief together with the existence of valves in the giraffe carotid arteries. Previous estimates of the giraffe heart was based on one report by Goetz and Keen (Goetz and Keen, 1957) and revealed a relative heart mass of 2.3% (Pedley et al, 1996). However, the estimate was based on only heart mass as no body mass was reported. The origin of the carotid valves "myth" is unknown; nonetheless, the valves are mentioned in the literature (Encyclopedia Britannia online and The Danish Encyclopedia e.g). We can now reject both "myths". Our study confirms that the giraffe heart constitute 0.5% of the body mass, which means that giraffes resemble other mammals in this respect. Though, not relatively bigger, the giraffe heart generates twice the pressure as any other mammalian hearts (Goetz and Budtz-Olsen, 1955; Van Citters et al, 1966; Brondum et al, 2009).

The relationship between volume of cardiomycytes and non-cardiomyocytic tissue in both young and adult giraffe LV was similar to the ratio of such cells in rats and humans (Bruel et al, 2007; Tang et al, 2009). *I.e.*, the percentage of muscular tissue in the giraffe LV is the same as in other mammals. The density of myocytic nuclei in the LV was significantly higher than in humans and rats (120×10^3 mm^{-3} in giraffes vs. 69.5×10^3 and 69×10^3 in rats and humans, respectively) (Bruel et al, 2002; Tang et al, 2009). Of greatest interest was the total number of nuclei in the adult LV, which was 62% higher than in the young LV. Proliferation of cardiomyocytes is heavily debated. We find that these results strongly indicate proliferation of myocytes after the post natal period. This could be an adaptation of the LV to overcome the increasing systemic afterload in the growing giraffes. For many years it has been thought that increasing LV wall stress due to e.g. hypertension is normalized only by parallel replication of sarcomeres and thereby thickening of each myocyte i.e. hypertrophic response (Karsner et al, 1925; Petersen and Baserga, 1965; Morkin and Ashford, 1968). It is likely, however, that the response to an increased afterload involves both hypertrophic and hyperplastic remodeling (Kajstura et al, 1994) and several studies have now shown that hyperplasia can take place after the postnatal period (Beltrami et al, 1995; Anversa and Kajstura, 1998; Beltrami et al, 2001; Leri et al, 2002; Engel, 2005; Bruel et al, 2007; Du et al, 2010; Mollova et al, 2013). We found the relative heart mass and the myocyte volume of the giraffe LV to be similar to that of other mammals including humans, but the total numbers of nuclei were significantly higher in the adult compared to the young LV. As studies indicate that an increase in the number of nuclei reflects an increase in myocyte number (Du et al, 2010), it seems reasonable to suggest that the significantly higher number of nuclei in the adult compared to the young indicates an increased number of myoctes (proliferation).

The average number of nuclei per myocyte was very high compared to other mammals, including humans (Bruel et al, 2007; Tang et al, 2009; Mollova et al, 2013). We speculate if this may optimize the protein and nucleic acid synthesis and hence allow for each myocyte to maintain proliferation. A probable mechanism suggested by Stephen et al, (2009) can be that the nuclei are "specialized" and play different roles. Stephen et al, (2009) hypothesized that one nucleus of a bi-nucleated myocyte may be dormant and only "activated" during cell replication, while the other nucleus is active in protein synthesis. The average number of nuclei in the giraffe myocyte was 4.2, which is considerably higher than any other known data from mammals and it is noteworthy that it coincides with the highest known mean arterial pressure.

The capillary volume of the giraffe LV is higher than what can be measured in human and rat hearts (5% versus 3.6 and 2.9%, in humans and rats, respectively) (Bruel et al, 2007; Tang et al, 2009). According to these results the oxygen delivery may be higher in the giraffe LV, either by a smaller diffusion distance or by a larger blood volume. This could be in accordance with a high work load of the heart to maintain the high mean arterial pressure, which would be interesting to investigate further.

The findings of the pronounced and abrupt narrowing of the tibial and median arteries supports earlier findings (Kimani et al, 1991), although our observation that the narrowing persist along the entire leg differ from Kimani's description (Kimani et al, 1991). The mean media to lumen ratio area increased more than six-fold within 2–4 cm. The tibial and median arteries are usually not considered to be regulative, but rather conduit vessels converting pulsatile flow into a continuous flow; therefore the finding of this sphincter-like structure was surprising. The findings that this area of the sudden narrowing of the tibial and median artery was richly innervated suggest it to be a pressure regulating mechanism. Kimani et al, 1991 also observed this local innervation of the artery. The ultrasound investigation of the median artery revealed constriction sometimes only in the spincther-like area and sometimes all along the leg distally of this area. Morphological analysis of the arterial wall showed the volume fraction of elastin fibers to be 38% above the narrowing and almost absent below the narrowing.

These findings do not point clearly to a true sphincter; however, they clearly indicate some kind of pressure regulating mechanism of the tibial and median artery. Moreover, that the artery constricts in response to norepinephrine and that a reduction of flow and a pressure difference between proximal and distal site of the narrowing is observed, strongly indicates a regulating mechanism (Petersen et al, submitted). The very high pressure measured in the legs of the giraffe was in accordance to previous measurements (Hargens et al, 1987). The pressure increase measured along the whole length of the artery (Ostergaard et al, 2011) was not significantly different from what was expected due to the influence of gravity and neither was the increase measured from just above the sphincter-like area to 20 cm proximal to the hoof (Petersen et al, submitted). At both measurements though, a smaller than expected pressure increase was observed (but not significant) and perhaps this could reflect a viscous pressure drop along the artery. This would be reasonable taking the structure of the sphincter-like area into account.

The very muscular area of the inferior caval vein near the kidney was not like any other vein morphology we are aware of. As figure 8 illustrates unusually thick longitudinal bundles of muscle exist in this area of the vein (Damkjær et al, in prep). The cross sectional area is large and at first impression of the gross anatomy of the vein it resembles the aorta. Just distal to the right atrium, the vein is "normal" and not either thick or muscular (Figure 8). Yet unpublished data indicate that the giraffe kidney can withstand twice the pressure as other mammalian kidneys, including humans (Damkjær et al, in prep). The pressure is high when the blood leaves the renal vein entering the caval vein. This is in agreement with the thick and muscular appearance of the caval vein in this area having to withstand a higher pressure than veins normally do.

The thick and muscular morphology of both the tibial and median artery as well as the inferior caval vein indicates a low systemic compliance below heart level in the giraffe. In contrast, the thickness and composition of the jugular vein resemble the classic appearance of veins; it is thin and compliant with low fraction of muscle tissue (Fig. 10) (Damkjær et al, in prep). It is noteworthy though, that the volume fraction of muscle tissue in the upper region of the jugular vein is significantly higher than that of the lower region (Fig. 10). This difference correlates well with the fact that blood accumulates in the upper region, when the head is lowered, whereas the lower region of the jugular probably never experience any notable pressure or stretching.

The overall morphological results point to a vascular system with a low compliance below heart level and a "normal" compliance above heart level. With a pressure twice as high as any other mammal and a low blood volume (Damkjær et al, in prep), it would be very reasonable for the function and protection of the entire cardiovascular system of the giraffe. Below heart level the protection against edemas and other tissue damage has to be high. Above heart level the venous structure must ensure venous distensibility to enable the accumulation of blood during lowering of the head (Brondum et al, 2009). This accumulation is thought to affect the MAP, and thereby influence how the gravitational pressure affects the cardiovascular system (Brondum et al, 2009).

Concluding Remarks

We have investigated whether "the myths" concerning the large heart and carotid valves could be verified, and found that they cannot. There are no valves in the carotid and the giraffe heart has a relative mass as other mammalian hearts, including the human.

We hypothesize that the unusually high number of cardiomyocyte nuclei of the giraffe LV could be an optimization of cardiomyocyte regeneration and replication or an optimized synthesis of protein and nucleic acid. A mechanism that ensures and maintains proliferation of myocytes, which our data strongly suggest exist in the giraffe LV during growth. We suggest proliferation of myocytes to be an adaptation to overcome the increasing systemic arterial afterload of the growing giraffe.

The sphincter-like area of the median and tibial artery is striking if these arteries are thought of as "simple" conduit arteries. However, the sudden change in morphology and innervation of this area indicates clearly a regulatory mechanism. The surprisingly muscular caval vein and the small media:lumen ratio of thetibial/median artery points to a low systemic compliance below heart level, which, together with the sphincter-like area may explain some of the protective mechanisms of the giraffe cardiovascular system.

A future goal would be a complete "mapping" of the giraffe cardiovascular system. But this encompasses a tremendous amount of work, tissue and animals. We have, however, contributed to this goal by investigating some of the important compartments of the system, which have enabled us to correlate both morphological and functional data. We have described several structural specializations that may explain some of the mechanisms working in a cardiovascular system exposed to the highest known pressure and hydrostatic pressure differences in any mammal.

Limitations

Animal Handling and Equipment

The ideal experiment is not possible; it would be what we have done plus a lot more, on free ranging and un-anesthetized animals; and intravascularly positioned devices should be known not to influence flow, pressure, blood coagulation etc.

We cannot ignore the influence of the anesthesia on the autonomic control (Matsukawa et al, 1993; Silverman and Muir, 1993; Constable, 1999; Maignan et al, 2000), though α-chloralose is known as an anesthetic providing stable heamodynamics (Silverman and Muir, 1993; Constable, 1999). Earlier studies were performed on sedated giraffes and it could be speculated whether this would be better than anesthetized giraffes. It may be the case from a physiological point of view; however, such experiments cannot fulfill the present ethical standards of animal experimentation.

In other words, we have planned the experiments carefully and as thorough as possible. We are aware, that there are unanswered questions, but we believe that we have lifted the veil from some of the mysteries regarding the giraffe's cardiovascular system.

Working with the giraffe as an experimental animal is a huge challenge with many limitations. The giraffe is extremely sensitive to anesthesia and both the digestive and the cardiovascular system are seriously challenged when the animal is placed horizontally. It requires unique facilities, professional veterinarians, many people for handling the anesthetized animal and specially designed probes, wires, data logger etc.

Tissue Handling

Tissue handling is also a huge challenge when working at outside facilities at a quarantine station in South Africa. Perfusion fixation of the tissue would have been preferred, but the facilities did not allow this. To avoid tissue shrinkage during preparation, plastic embedding was preferred. However, this

31

limited our staining possibilities along with the fact that no antibodies specific for giraffe tissue have been developed. Unfortunately, the PAS-M staining visualizing the intercalated discs and thereby enabling differentiation of each cardiomyocyte, did not stain the young tissue. This made the collecting of number of nuclei per myocyte on young tissue impossible. The circumstances of the outdoor setup in the South African outback forced us to fixate all tissue in buffered formalin, and made it impossible to perform microstructural and molecular studies needing fresh tissue and a well-equipped laboratory.

Stereology

Stereology is a very efficient method for gaining design-unbiased data, but it also has its drawbacks. Using stereology, 100% of the organ of interest has to be available in order to be design-unbiased. In a study like this, it complicates the tissue collection as we needed to fixate entire giraffe hearts to bring back to the laboratory as it is difficult to sample properly in the outback. Stereology is time consuming, which limits the use of the methods in some studies/diagnostics. Tissue shrinkage and deformation are important to consider when preparing tissue samples, as these factors can have great influence on measurements of volume and densities. To limit the degree of shrinkage, plastic embedment was used (glycolmethacrylate)(Dorph-Petersen et al, 2001). To ensure whether correction for shrinkage was necessary, the tissue volume was determined before and after embedding of the tissue.

Perspectives

We have had the opportunity to study the giraffe cardiovascular system, which have enabled us to correlate the findings of morphology of arteries, veins and heart to the functional parameters of the vascular system. The findings have revealed unknown fundamental mechanisms and structures. We speculate that the sphincter-like area of the median and tibial artery is a mechanism of vascular protection; however we need further confirmation of this in free ranging giraffes. Thus, for future research it would be very appealing to develop methods and approaches to measure pressure in upper and lower leg simultaneously, in free ranging, un-anesthetized giraffes. It is also of great interest to determine whether similar mechanisms exist in other long necked animals.

During the morphological study of the heart, some questions have arisen that could have been answered if ultrastructural studies of the cardiomyocytes had been possible. However, as described in the limitation section, the experimental setup required to handle animals as large as the giraffe, limits the possibilities of collecting tissue for ultrastructural studies. During the last expedition though, we collected frozen tissue samples for proteomics. The analysis will be launched later this year. Perhaps studying the proteomics of the giraffe cardiomyocytes will reveal the function of the many nuclei, as exactly the many nuclei per myocyte, indeed raises new questions. It is noteworthy that the very high number of nuclei per cardiomyocyte coincides with the highest mean arterial pressure known in any living mammal.

In conclusion, the present study have together with the DaGiR team, examined several structural and functional parameters and our knowledge of vascular morphology and regulatory mechanisms has been extended.

Summary

Being the tallest living animal, giraffes are endowed with a mean arterial pressure (MAP) twice as high as other mammals, including humans. Also the hydrostatic pressure differences within the giraffe cardiovascular system are enormous. This undoubtedly causes serious challenges to the cardiovascular system of the giraffe. Mechanisms probably have evolved to protect the cardiovascular system from the adverse problems of a high MAP. Previous it was proposed that the giraffe do not develop heart failure because the size of the heart twice that of other mammals. However, we now know that the giraffe heart constitutes 0.5% of body mass just as in other mammals. Moreover the systemic circulation must be challenged by the very high MAP.

The aims of the present thesis are to unveil some of the mechanisms underlying function and protection of the giraffe cardiovascular system. We have sought to extend our knowledge of the adaptaions system and to confirm or reject some of the myths that inevitably exist in connection with a mammal reaching 5–6 meters in height. The primary focus has been on the morphology of the heart, the conduit arteries of the lower extremeties, the neck and abdominal large veins (jugular and inferior caval vein) and the carotid artery.

To study the cardiovascular morphology, heart, arteries, veins and much more were collected for examination. Before euthanasia and autopsy, physiological measurements were performed either under anesthesia or in free-ranging giraffes. The main part of the results has been gained by stereological methods.

The volume fraction, total volume, density and total numbers of cardiomyocytes and non-cardiomyocytes of the left ventricle (LV) were estimated. Further, the number of nuclei per myocyte and the volume fraction of capillaries were estimated.

In the lower extremities, the lumen- and tunica media area of the tibial and median artery were measured as was the ratio between these. The volume fraction of elastin and the innervation of these vessels were estimated. These data were followed up with pressure measurements and ultrasound imaging of the median artery. The carotid artery was examined and the morphology of large neck and abdominal veins were investigated.

The main findings revealed a significantly higher total number of myocyte nuclei in the adult LV compared to the newborn/young LV. The number of nuclei per myocyte were surprisingly high compared to other mammalian myocyteswhich may function as an optimization of cardiomyocyte regeneration and replication.

An abrupt and significant decrease of the tibial and median artery lumen area and of the media:lumen ratio was observed. In addition, the innervation was significantly higher in this area. Together these facts points to a regulatory mechanism. To our knowledge this is not previously described in arteries considered conduit arteries.

The caval vein was unusuallythick and muscular near the renal vein, compared to other locations of the vein, suggesting the ability to withstand a high venous pressure. In contrast the jugular veins were thin and compliant as expected. The appearance of arteries and veins investigated points to a system with a low compliance below heart level and a "normal" compliance above heart level, which would be advantageously for function and protection in a cardiovascular system exposed to high pressure differences and a low blood volume.

Resumé (Summary in Danish)

Med sin usædvanlige højde og som det højeste nulevende dyr, er giraffen udstyret med et arterielt middel blodtryk dobbelt så højt som andre pattedyr, inklusiv mennesket. Også de hydrostatiske tryk forskelle i giraffens kredsløb er enorme. Dette medfører utvivlsomt store udfordringer for udviklingen af giraffens kredsløb. Beskyttende mekanismer må være udviklet I giraffens kredsløb, for at imødekomme de alvorlige konsekvenser et højt arterielt blodtryk kan have. Tidligere er det foreslået, at den relative masse af giraffens hjerte var dobbelt så stor som andre pattedyrs hjerter, men seneste studier viser, at giraffens hjerte har en relative størrelse på 0.5%, ligesom hos andre pattedyr.

Endvidere må også det systemiske kredsløb have fordret en målrettet udvikling for at imødekomme det høje tryk.

Formålet med disse studier, har været at belyse dele af de funkionelle og beskyttende mekanismer i giraffens kredsløb. Studierne forsøger, at belyse tilpasninger i giraffens kredsløb og be- eller afkræfte nogle af de myter som uundgåeligt eksisterer om dette ejendommelige dyr.

Morfologien af hjerte, tilførselsarterier i ekstremiteterne, store hals- og abdominale vener samt karotid arterierne har været det primære fokus i disse studier.

Hjerter, arterier, vener m.m. blev dissekeret og indsamlet til morfologiske studier. Forskellige fysiologiske målinger i både fritgående og bedøvede giraffer blev udført inden aflivning og autopsi.

Hovedparten af resultaterne er opnået ved hjælp af stereologiske metoder.

Volumenfraktion, den total volumen, tætheden og det total antal kardiomyocytter og ikke-kardiomyocytter i den venstre hjerteventrikel (LV) blev målt. Endvidere blev antallet af cellekerner per kardiomyocyt og volumefraktionen af kapillærer estimeret i LV.

I underekstremiteterne blev areal af karlumen og tunica media langs a. tibialis målt samt ratio mellem disse. Også volumenfraktion af elastin og inneveringen af karret blev estimeret. Disse data blev fulgt op med en tryk profil fra hele benlængden. Karotid arterier blev undersøgt og morfologien af hals- og abdominal vener blev analyseret.

Resultaterne viste et signifikant højere antal kardiomyocyt cellekerner i den voksne venstre ventrikel sammenlignet med antallet i nyfødte og unge dyrs ventrikler. Antallet af cellekerner per kardiomyocyt var højt sammenlignet med andre pattedyr, og vi foreslår, at dette kan være en tilpasning til optimering af kardiomyocyt regeneration og replikation.

En pludselig og signifikant indsnævring af det arterielle lumen og markant fald i media:lumen areal forhold, blev fundet. Yderligereøgedes inner-veringen af karvæggen signifikant i dette område. Disse observationer under støtter eksistensen af en regulatorisk mekanisme. Dette er, os bekendt, ikke beskrevet før i store tilførselskar, som ikke normalt betragtes som regulatoriske.

Vena cava var usædvanlig kraftig og muskuløs nær vena renalis sam-menlignet med andre lokationer af venen, hvilket kunne indikere, at venen modstår et højt venøst tryk. I modsætning til dette, var jugular venerne tynde og eftergivelige, som forventet.

De undersøgte kar peger på et systemisk kredsløb med en lav eftergive-lighed under hjerte niveau og en "normal" eftergivelighed over, hvilket vil være fordelagtigt i et systemisk kredsløb, som er udsat for store tryk forskelle og lav blodvolumen.

Bibliography

1 Anversa P, Kajstura J (1998). Ventricular myocytes are not terminally differentiated in the adult mammalian heart. Circulation Research 83:1–14.

2 Badeer HS, Hicks JW (1992). Hemodynamics of vascular waterfall - Is the analogy justified. Respiration Physiology 87:205–217.

3 Barri Y (2008). Hypertension and kidney disease: A deadly connection. Current Science Inc 10: 39–45.

4 Beltrami AP, Urbanek K, Kajstura J, Yan SM, Finato N, Bussani R, Nadal-Ginard B, Silvestri F, Leri A, Beltrami CA, Anversa P (2001). Evidence that human cardiac myocytes divide after myocardial infarction. New England Journal of Medicine 344:1750–1757.

5 Beltrami CA, Finato N, Rocco M, Feruglio GA, Puricelli C, Cigola E, Sonnenblick EH, Olivetti G, Anversa P (1995). The cellular casis of cilated cardiomyopathy in humans. Journal of Molecular and Cellular Cardiology 27:291–305.

6 Brændgaard H, Gundersen HJG (1986). The impact of recent stereological advances on quantitative studies of the nervous system. Journal of Neuroscience Methods 18: 39–78.

7 Brondum E, Hasenkam JM, Secher NH, Bertelsen MF, Grondahl C, Petersen KK, Buhl R, Aalkjaer C, Baandrup U, Nygaard H, Smerup M, Stegmann F, Sloth E, Ostergaard KH, Nissen P, Runge M, Pitsillides K, Wang T (2009). Jugular venous pooling during lowering of the head affects blood pressure of the anesthetized giraffe. American Journal of Physiology-Regulatory Integrative and Comparative Physiology 297: R1058–R1065.

8 Brook BS, Pedley TJ (2002). A model for time-dependent flow in (giraffe jugular) veins: uniform tube properties. Journal of Biomechanics 35:95–107.

9 Bruel A, Christoffersen TEH, Nyengaard JR (2007). Growth hormone increases the proliferation of existing cardiac myocytes and the total number of cardiac myocytes in the rat heart. Cardiovascular Research 76:400–408.

10 Bruel A, Oxlund H, Nyengaard JR (2002). Growth hormone increases the total number of cardiac myocyte nuclei in young rats but not in old rats. Mechanisms of Ageing and Development 123:1353–1362.

11 Constable PD (1999). Acute endotoxemia increases left ventricular contractility and diastolic stiffness in calves. Shock 12: 391–401.

12 Damkjær M, Funder J, Smerup M, Østergaard K, Brondum ET, Wang T, Baandrup U, Hørlyck A, Hasenkam JM, Bertelsen MF, Markussen N, Chemnitz J, Danielsen CC, Grøndahl C, Candy G. and Bie P. High renal interstitial hydrostatic pressure and low glomerular filtration rate in the giraffe (Giraffacamelopardalis). In preparation.

13 Dorph-Petersen KA, Nyengaard JR, Gundersen HJG (2001). Tissue shrinkage and unbiased stereological estimation of particle number and size. Journal of Microscopy-Oxford 204:232–246.

14 Du Y, Plante E, Janicki JS, Bower GL (2010). Temporal evaluation of cardiac myocyte hypertrophy and hyperplasia in male rats secondary to chronic volume overload. American Journal of Pathology 177: 1155–1163.

15 Engel FB (2005). Cardiomyocyte proliferation - A platform for mammalian cardiac repair. Cell Cycle 4:1360–1363.

16 Goetz RH, Budtz-Olsen O. (1955). Scientific Safari; The circulation of the giraffe. South Africa Medicin Journal 29: 773–776.

17 Goetz RH, Keen EN (1957). Some aspects of the cardiovascular system of the giraffe. Angiology 8: 542–564.

18 Goetz RH, Warren JV, Gauer OH, Patterson JL, Doyle JT, Keen EN, Mcgregor M (1960). Circulation of the giraffe. Circulation Research 8: 1049–1058.

19 Gundersen HJG, Bagger P, Bendtsen TF, Evans SM, Korbo L, Marcussen N, Moller A, Nielsen K, Nyengaard JR, Pakkenberg B, Sorensen FB, Vesterby A, West MJ (1988). The new stereological tools - disector, fractionator, nucleator and point sampled intercepts and their use in pathological research and diagnosis. Apmis 96:857–881.

20 Gundersen HJG, Jensen EB (1987). The Efficiency of Systematic-Sampling in Stereology and Its Prediction. Journal of Microscopy-Oxford 147:229–263.

21 Hargens AR, Millard RW, Pettersson K, Johansen K (1987). Gravitational hemodynamics and edema prevention in the giraffe. Nature 329:59–60.

22 Hicks JW, Badeer HS (1989). Siphon mechanism in collapsible tubes - Application to circulation of the giraffe head. American Journal of Physiology 256:R567-R571.

23 Kajstura J, Zhang X, Reiss K, Szoke E, Li P, Lagrasta C, Cheng W, Darzynkiewicz Z, Olivetti G, Anversa P (1994). Myocyte cellular hyperplasia and myocyte cellular hypertrophy contribute to chronic ventricular remodeling in coronary-Artery narrowing-induced cardiomyopathy in rats. Circulation Research 74:383–400.

24 Karsner HT, Saphir O, Todd TW (1925). The state of the cardiac muscle in hypertrophy and atrophy. American Journal of Pathology 1:351–372.

25 Kimani JK, Mbuva RN, Kinyamu RM (1991). sympathetic innervation of the hindlimb arterial system in the giraffe (Giraffa-Camelopardalis). Anatomical Record 229: 103–108.

26 Korsgaard N, Aalkjaer C, Heagerty AM, Izzard AS, Mulvany MJ (1993). Histology of subcutaneous small arteries from patients with essential-hypertension. Hypertension 22:523–526.

27 Schoel FJ and Mitchell RN; The Heart in Kumar Vet al. (eds): Robbins Cotran Pathologic basis of diseases. 8th Edition. Philadephia, *Elsevier and Saunders*, 2005, p. 529–530.

28 Leri A, Kajstura J, Anversa P (2002). Myocyte proliferation and ventricular remodeling. Journal of Cardiac Failure 8:S518-S525.

29 Maignan E, Dong WX, Legrand M, Safar M, Cuche JL (2000). Sympathetic activity in the rat: effects of anaesthesia on noradrenaline kinetics. Journal of the Autonomic Nervous System 80:46–51.

30 Matsukawa K, Ninomiya I, Nishiura N (1993). Effects of anesthesia on cardiac and renal sympathetic nerve activities and plasma catecholamines. American Journal of Physiology - Regulatory, Integrative and Comparative Physiology 265:R792-R797.

31 Mitchell G, Maloney SK, Mitchell D, Keegan DJ (2006). The origin of mean arterial and jugular venous blood pressures in giraffes. Journal of Experimental Biology 209: 2515–2524.

32 Mitchell G, Skinner JD (2009). An allometric analysis of the giraffe cardiovascular system. Comparative Biochemistry and Physiology A-Molecular & Integrative Physiology 154:523–529.

33 Mollova M, Bersell K, Walsh S, Savla J, Das LT, Park SY, Silberstein LE, dos Remedios CG, Graham D, Colan S, Kühn B (2013). Cardiomyocyte proliferation contributes to heart growth in young humans. Proceedings of the National Academy of Sciences 110: 1446–1451.

34 Morkin E, Ashford TP (1968). Myocardial dna synthesis in experimental cardiac hypertrophy. American Journal of Physiology 215:1409–1413.

35 Mulvany MJ (1998). Vasuclar remodeling of resistance vessels: Can we define this? Cardiovascular Research 41:9–13.

36 Mulvany MJ (2002). Small artery remodeling and significance in the development of hypertension. News in Physiological Sciences 17:105–109.

37 Nyengaard JR, Gundersen HJG (1992). The Isector - A simple and direct method for generating isotropic, uniform random sections from small specimens. Journal of Microscopy-Oxford 165:427–431.

38 Nyengaard JR (1999). Stereologic methods and their application in kidney research. Journal of the American Society of Nephrology 10:1100–1123.

39 Ostergaard K, Bertelsen M, Brondum E, Aalkjaer C, Hasenkam J, Smerup M, Wang T, Nyengaard J, Baandrup U (2011). Pressure profile and morphology of the arteries along the giraffe limb. Journal of Comparative Physiology B: Biochemical, Systemic, and Environmental Physiology 181:691–698.

40 Pedley TJ, Brook BS, Seymour RS (1996). Blood pressure and flow rate in the giraffe jugular vein. Philosophical Transactions of the Royal Society of London Series B-Biological Sciences 351:855–866.

41 Petersen KK, Hørlyck A, Østergaard KH, Andresen J, Brøgger T, Skovgaard N, Telinius N, Laher I, Bertelsen, MF, Grøndahl C, Smerup M, Secher NH, Brøndum E, Hasenkam JM, Wang T, Baandrup U, Aalkjær C. Protection against high intravascular pressure in the leg of giraffes. Submitted to American Journalof Physiology, March 2013.

42 Petersen RO, Baserga R (1965). Nucleic acid and protein synthesis in cardiac muscle of growing and adult mice. Experimental Cell Research 40:340–352.

43 Seymour RS, Blaylock A (2000). The principle of Laplace and scaling of blood pressure and ventricular wall stress in mammals and birds. Physiological and Biochemical Zoology 73;389–405.

44 Silverman J, Muir WW (1993). A review of laboratory animal anesthesia with chloral hydrate and chloralose. Laboratory Animal Science 43:210–216.

45 Simmons RE, Altwegg R (2010). Necks-for-sex or competing browsers? A critique of ideas on the evolution of giraffe. Journal of Zoology 282:6–12.

46 Stephen MJ, Poindexter BJ, Moolman JA, Sheikh-Hamad D, Bick RJ (2009). Do binucleate cardiomyocyte have a role in myocardial repair? Insight using isolated rodent myocytes and cell culture. Open Cardiovascular Medicin Journal 3:1–7.

47 Tang Y, Nyengaard JR, Andersen JB, Baandrup U, Gundersen HJ (2009). The application of stereological methods for estimating structural parameters in the human heart. Anatomical Records 292:1630–1647.

48 Van Citters RL, Kemper WS, Franklin DL (1966). Blood pressure responses of wild giraffes studied by radio telemetry. Science 152:384–386.

49 Williams JR, Vogler NJ, Kilo C (1971). Regional variations in width of basement membrane of muscle capillaries in man and giraffe. American Journal of Pathology 63:359–370.

www.ingramcontent.com/pod-product-compliance
Lightning Source LLC
Chambersburg PA
CBHW061840220326
41599CB00027B/5360